THINKING ABOUT SCIEN

I0000367

Our Solar System

Written by Rebecca Stark

The purchase of this book entitles the individual teacher to reproduce copies of the student pages for use in his or her classroom exclusively. The reproduction of any part of the work for an entire school or school system or for commercial use is prohibited.

ISBN 978-1-56644-050-9

© 2016 Barbara M. Peller

Educational Books 'n' Bingo

Previously published by Educational Impressions, Inc.

Printed in the U.S.A.

Table of Contents

To the Teacher

Our Solar System is a comprehensive, fun-filled unit on the solar system. Students learn important facts about the history of astronomy, the sun, the planets, our moon, asteroids, meteoroids, comets, and more. They are also presented with opportunities to practice crucial critical- and creative-thinking skills. A variety of types of activities are included: creative writing, experimenting, research, math, analyzing, evaluating, and more.

A fun What's the Question? game is provided to reinforce the concepts learned in the unit. In addition, a crossword puzzle is included. The puzzle may be used to evaluate knowledge or just for fun!

Our Solar System: What Is It?

Our solar system is made up of a star, which we call our sun, and the objects which revolve around it. Most important among those objects are the nine planets and their moons. Also part of the solar system, however, are smaller objects, such as asteroids, comets, meteoroids, dust, and gases.

The four planets closest to the sun are called the inner planets. In order of their distance from the sun, the inner planets are Mercury, Venus, Earth, and Mars. The four outer planets are much farther away. Those planets are Jupiter, Saturn, Uranus, and Neptune.

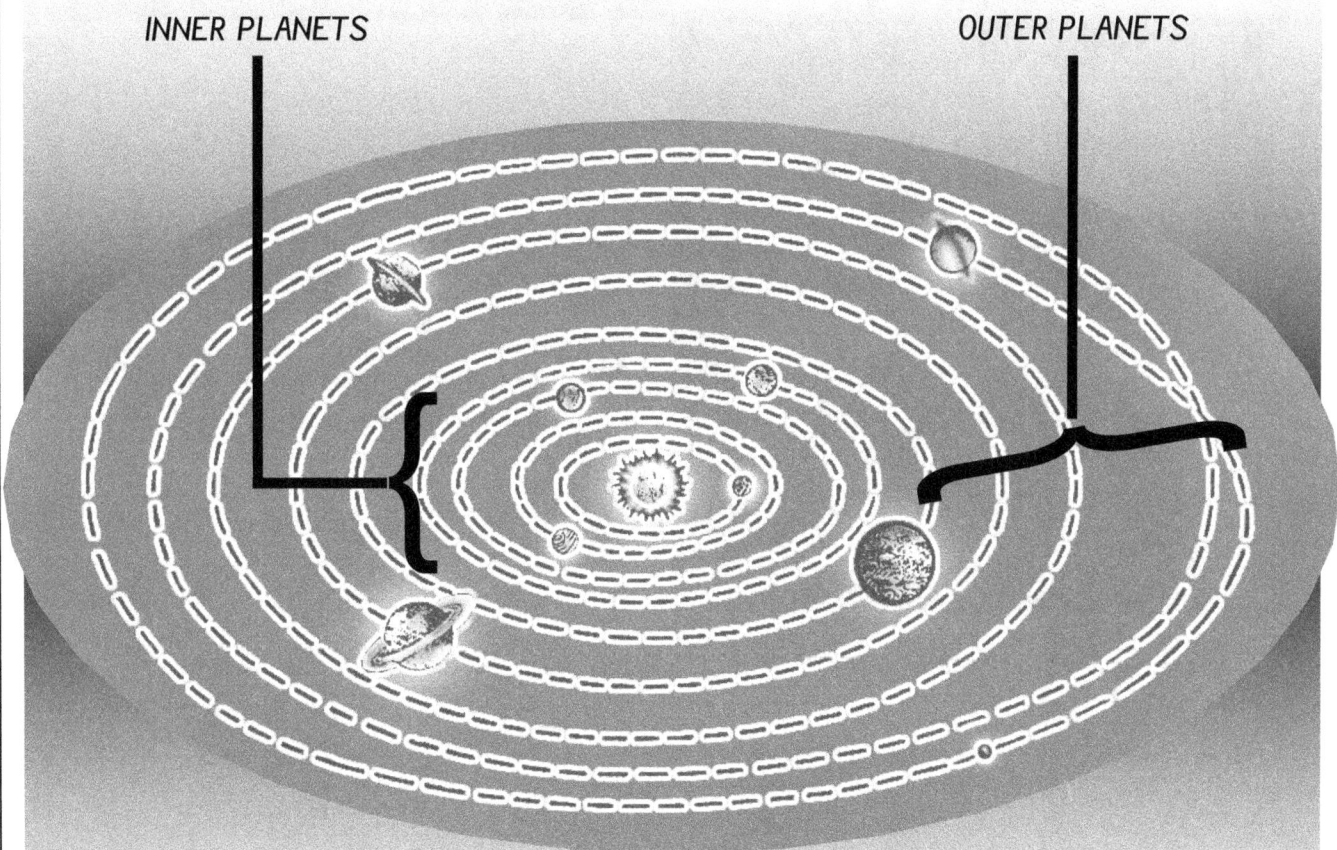

INNER PLANETS   OUTER PLANETS

© Educational Impressions, Inc.

Ancient Views

In A.D. 140 Greek astronomer Ptolemy put forth a view of the universe that was widely accepted as fact for about 1,500 years. Research Ptolemy and draw a chart that depicts Ptolemy's view of the universe.

Polish astronomer Nicolaus Copernicus, who lived from 1473–1543, set forth a heliocentric theory of the universe. It was in opposition to the widely accepted views of Aristotle and Ptolemy.

Define the terms "geocentric" and "heliocentric" as they apply to the above theories.

_____

_____

_____

_____

_____

_____

© Educational Impressions, Inc.

Astronomical Match-up

Research the people on the left. Then match them with the descriptions on the right.

_____ 1. ARISTARCHUS

_____ 2. TYCHO BRAHE

_____ 3. COPERNICUS

_____ 4. GALILEO GALILEI

_____ 5. HIPPARCHUS

_____ 6. JOHANNES KEPLER

_____ 7. ISAAC NEWTON

_____ 8. PTOLEMY

_____ 9. THALES

S. Danish astronomer (1546–1601) who held a compromise theory; he believed that the planets orbit the sun, but he also believed that the sun orbits Earth.

R. Italian scientist (1564–1642) who, although he didn't invent the telescope, did build a refracting one; he discovered the four large moons of Jupiter.

A. Third-century B.C. Greek astronomer whose early sun-centered views of the universe were ignored.

T. Polish astronomer (1473–1543) who published a sun-centered theory of the universe; his theory was wrong in that he also believed the stars orbited the sun.

N. German mathematician (1571–1630) whose laws of planetary motion, including elliptical orbits, are the basis for our modern views.

O_1. Second-century Greek astronomer who catalogued 850 stars, which he classified according to their brightness.

O_2. English mathematician and scientist (1642–1727) who formulated the theory of universal gravitation.

M. In _Almagest,_ this 2nd-century A.D. Greek astronomer summarized previous Greek thoughts, producing a detailed description of the celestial bodies' orbits around Earth.

Y. Greek philosopher (640?–546 B.C.); he was a keen observer of the sky and predicted an eclipse in 585 B.C.

If you matched the columns correctly, your answers should spell the word that means "the scientific study of the universe beyond the Earth."

_____

Create an Astronomy Time Line

Use what you learned in the Astronomical Match-up activity to create an astronomy time line. Do some outside research and add any people and events you think would enhance your time line.

An Early History of Astronomy

© Educational Impressions, Inc.

Just an Average Star!

Our sun is just an average star—one of billions of stars in the universe. Like other stars, our sun is a huge, hot glowing ball of gases, mostly hydrogen and helium. At the sun's surface, or photosphere, the temperature goes to about 10,000°F (5,500°C). At its core, the temperature reaches about 27,000,000°F (15,000,000°C).

The most important thing that makes a celestial body a star is the fact that it is luminous. In other words, it gives off self-generated light. Planets and other celestial bodies shine only because of reflected light.

Our sun is classified as a dwarf star. That means that it has relatively low mass and average or below average luminosity, or the quality of emitting self-generating light. Highly luminous and massive stars are classified as giants. Some giants are 100,000 times as luminous as our sun. Some very faint stars, called white dwarfs, are 1,000 times less luminous than our sun.

At the core, or center, of the sun there are constant nuclear explosions. These explosions turn the hydrogen into helium. Some of the hydrogen gas turns into invisible X-ray energy; this energy is released as heat and light. It takes millions of years for the X-rays to travel from the core, through the radioactive and convective zones, to the surface. When it finally reaches the surface, however, it takes only about eight and a half minutes to reach Earth. Although the sun loses over 4 million tons of hydrogen every second, we don't have to worry about its running out. It has enough hydrogen to shine for another 5 billion years!

Like Earth, the sun, too, has an atmosphere. Its inner atmosphere is called the chromosphere; it is a thin ring about 6,200 miles (10,000 kilometers) thick. *Chromo* means "color." Although we usually can't see it, during a total solar eclipse it becomes visible, giving off a reddish glow. Also seen during a total solar eclipse (or with a cromograph) is the corona, the outermost atmosphere of the sun. The corona extends several million miles into space and appears as a faintly colored luminous ring, or halo. Temperatures in the corona reach as high as 3,000,000°F (1,440,000°C).

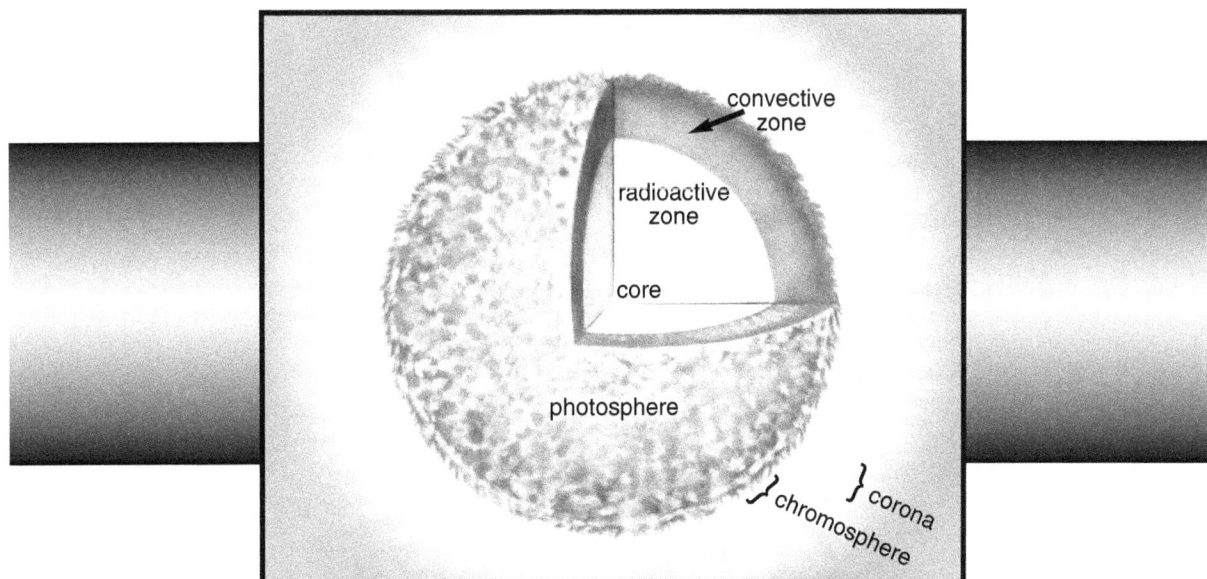

Activities

A solar eclipse occurs when the moon comes between Earth and the sun. Draw a diagram that illustrates a solar eclipse.

Create a safety poster warning people never to look directly at the sun. Include a diagram of a safe way to view the sun; for example, in order to view a solar eclipse.

Define photosynthesis. Use photosynthesis to explain why all life on Earth depends upon the light from the sun.

Find out what is meant by an astronomical unit, or AU.

The intense magnetic fields in the sun's interior are responsible for many changes in the physical structure of the photosphere. Define the following phenomena known collectively as solar activity: sunspots, faculae, and flares.

© **Educational Impressions, Inc.**

About the Planets

A planet is a body that directly orbits a star and does not shine by its own light. Most of the planets in our solar system—those that revolve around our sun—have at least one satellite, or moon. Mercury and Venus are the only ones that do not have any moons.

There are eight planets in our solar system. In addition to Earth, only Mercury, Venus, Mars, Jupiter, and Saturn were known to the ancients. The others weren't discovered until astronomers began using telescopes for their observations. Uranus was discovered in 1781; and Neptune was discovered in 1846. Pluto, which was classified as the ninth planet in our solar system for 76 years, was discovered in 1930. For a long time many astronomers also believed that a tenth planet, referred to as Planet X, might exist.

The inner planets—those closest to the sun—have many characteristics in common. Included are Mercury, Venus, Earth, and Mars. Because of their many similarities to Earth, they are sometimes called the terrestrial planets. All four terrestrial planets have hard crusts, composed of lightweight silicate rocks. Rocky mantles of heavier rock lie beneath the crusts. Heavy iron cores form their centers.

Four of the outer planets—Jupiter, Saturn, Uranus, and Neptune—are sometimes called the giant gas planets. Because these planets are so far from the sun, they did not lose great amounts of lightweight gases such as hydrogen, the way the inner planets did. They became much more massive, therefore, than the terrestrial planets. The atmospheres of the giant gas planets are composed mostly of hydrogen. It is, of course, very cold on these planets. Neptune, for example, has a temperature at its surface of about -360°F (-218°C).

© Educational Impressions, Inc.

Planets Fact Sheet

IN ORDER OF DISTANCE FROM THE SUN

Mercury
Venus
Earth
Mars
Jupiter
Saturn
Uranus
Neptune

IN SIZE ORDER: SMALLEST TO LARGEST

Mercury
Mars
Venus
Earth
Neptune
Uranus
Saturn
Jupiter

NUMBER OF KNOWN MOONS *

Mercury: 0
Venus: 0
Earth: 1
Mars: 2
Jupiter: 63*
Saturn: 60*
Uranus: 27*
Neptune: 13*

*It is very possible that new moons will be discovered in the future.

APPROXIMATE ORBITING TIME IN EARTH–DAYS/YEARS

Mercury: 87.97 Earth-Days
Venus: 224.7 Earth-Days
Earth: 365.24 Earth-Days
Mars: 687 Earth-Days
Jupiter: 11.86 Earth-Years
Saturn: 29.42 Earth-Years
Uranus: 84.01 Earth-Years
Neptune: 164.7 Earth-Years

APPROXIMATE ROTATION TIME IN EARTH–HOURS/DAYS

Mercury: 58.65 Earth-Days
Venus: 243 Earth-Days
Earth: 24 Earth-Hours (1 Day)
Mars: 24.6 Earth-Hours
Jupiter: 9.9 Earth-Hours
Saturn: 10.5 Earth-Hours
Uranus: 17.24 Earth-Hours
Neptune: 17.2 Earth-Hours

© Educational Impressions, Inc.

Activities

Draw a chart that explains what causes daytime and nighttime on the planets.

Create a Venn diagram to compare and contrast a star and a planet.

Use the facts given on the Planets Fact Sheet to create three What Planet Am I? riddles.

Create a mneumonic phrase or sentence (or another memory device) to help people remember the planets in size order.

Create a mneumonic phrase or sentence (or another memory device) to help people remember the planets in order of their distance from the sun.

© Educational Impressions, Inc.

Mercury

At an average of 36 million miles away, Mercury is the closest to the sun. Because it is so close to the sun, the planet's surface is very, very hot! In some spots, it reaches about 800°F (438°C). Nighttime temperatures, however, quickly drop to as low as -300°F (-184° C).

The closer a planet is to the sun, the stronger the pull of the sun's gravity and the faster the planet revolves around the sun. Of course, it also has a shorter distance to travel. Mercury takes about 88 Earth-days to make one complete revolution around the sun.

Activities

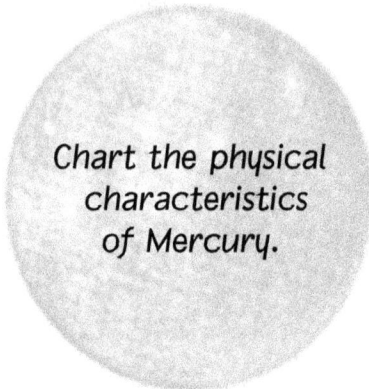

Chart the physical characteristics of Mercury.

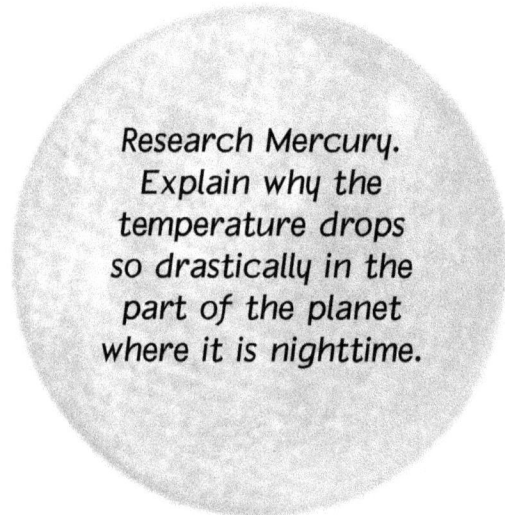

Research Mercury. Explain why the temperature drops so drastically in the part of the planet where it is nighttime.

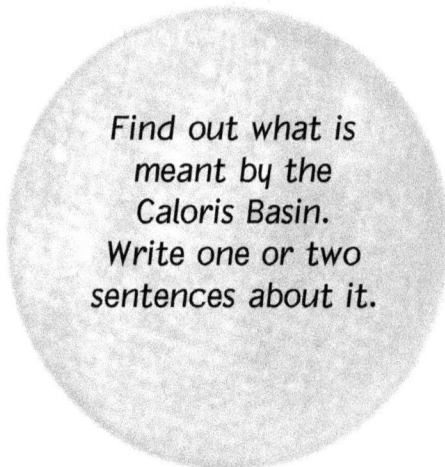

Find out what is meant by the Caloris Basin. Write one or two sentences about it.

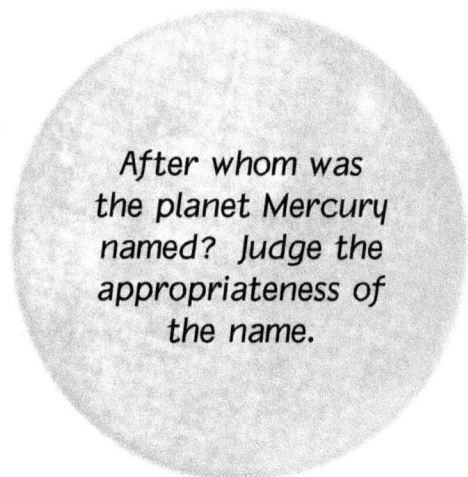

After whom was the planet Mercury named? Judge the appropriateness of the name.

© **Educational Impressions, Inc.**

Venus

Venus is the second planet from the sun. It is one of the four inner planets, called the terrestrial planets. Venus has about the same diameter, mass, and rock composition as Earth; however, Venus has no plate tectonics.

Temperatures on the surface of Venus are much, much higher than those on Earth. They are even higher than temperatures on the daytime side of Mercury, which is closer to the sun! Temperatures on Venus reach 900°F (480° C). The extremely high temperatures are because of it has a very dense atmosphere.

Because Venus is tilted on its axis at a 177.3° angle, (Earth is tilted at 23.45°), the planet rotates in a "backward" direction compared to Earth and the other planets.

Activities

Use a Venn diagram to chart the similarities and differences between Earth and Venus.

Use the greenhouse effect to explain the very high surface temperatures on Venus.

Define plate tectonics.

Review the list of rotation times and revolution times on the Planets Fact Sheet. In what way is Venus unique?

On Earth the sun rises in the east and sets in the west. Explain why it is opposite on Venus.

Judge the choice of the name Venus for this planet.

Earth

The third planet from the sun—and to those of us who live here, the most important planet in the solar system—is Earth! To the best of our knowledge, it is, in fact, the only planet that has life.

Earth is one of the terrestrial planets. In fact, the term "terrestrial" comes from the Latin *terra,* meaning "earth." Like the other terrestrial planets, Earth has different layers of rocky materials. The top layer, or crust, is very thin. Most of the internal activity occurs in the next layer down, called the mantle, which is 1,800 miles (2,900 kilometers) thick. Beneath the mantle is a liquid core and then an inner, solid core.

More than seventy percent of Earth's surface is covered with water. Earth's solid outer layer, the lithosphere, comprises the crust and the very top part of the mantle. It is divided into twelve large, rigid plates. All the plates contain a continent and a segment of ocean floor with the exception of the Pacific plate, which has no continent. The plates drift. When they collide or move apart, earthquakes or volcanoes may occur.

Earth has one satellite, or moon, orbiting it. Although Earth's moon is smaller than Earth—about one quarter its size—it is large enough to cause some astronomers to consider Earth and its moon a double planet.

Activities

Pretend that you are an astronaut. Describe in a letter to your family how Earth appeared from space.

Make a cutaway drawing of Earth. Show the crust, mantle, and core.

Demonstrate with a globe and a flashlight how the tilt of Earth's axis causes the seasons.

Find out what is meant by Pangaea.

Draw a picture that shows why Earth is unique in the solar system.

© Educational Impressions, Inc.

What a Planet!

When *Voyager* space probes were launched in 1977, scientists attached gold-coated photograph records to them. This was done because of the *very* remote possibility that an extraterrestrial civilization would ever recover them. Included are encoded pictures of Earth and human beings, greetings in 54 languages, sounds of Earth, and several musical selections.

Write a description of Earth from the point of view of an extraterrestrial being who has decoded the phonograph recording.

A Planet Called Earth

_____

_____

_____

_____

_____

_____

_____

_____

_____

_____

_____

_____

_____

_____

Mars

Mars, the fourth planet from the sun, has always been of interest to us. Sometimes Mars is called the Red Planet because its rock-strewn surface is covered with iron oxide, a rusty substance. It was this reddish color of the dust that inspired the ancients to name it after Mars, the Roman god of war.

The surface of Mars has many craters, huge canyons, and towering volcanoes. The gigantic canyon system known as Valles Marineris makes the Grand Canyon look like a mere ditch! Olympus Mons, an extinct volcano, is three times larger than Mauna Loa, Earth's largest volcano. It is also almost three times as high as Mount Everest, the highest mountain on Earth. It is, in fact, the highest mountain in our solar system.

People have long wondered if life exists or has ever existed on Mars. Two elements necessary for life as we know it are water and nitrogen. At one time liquid water did exist on Mars. The two polar caps, once thought to be only frozen carbon dioxide (dry ice) are now known to be partly water ice. The atmosphere, although mostly carbon dioxide, does contain almost 3 percent nitrogen. So far, however, no evidence of organic material of any kind has been found. Most scientists believe that no life exists near the surface of Mars. It is possible, however, that life in some simple form might exist in the wetter environments beneath the surface.

Activities

The moons of Mars are named Phobos and Deimos. Find out why those names were chosen.

Although it has many geological features in common with Earth, Mars has no plate tectonics. Use this fact to explain the enormous size of many geological features on Mars.

Pretend that you have been asked to take part in a mission to Mars. Write a letter to NASA in which you accept or decline the invitation.

Create a profile for the perfect candidate to be the first human to set foot on Mars.

Create a travel brochure enticing people to visit Olympus Mons or the Valles Marineris.

© **Educational Impressions, Inc.**

Jupiter

Jupiter, the fifth planet from the sun, is the largest planet in our solar system. All of the other planets would fit inside it! With an approximate diameter of 86,000 miles (138,150 kilometers), it is the second largest body in our solar system. Of course, the sun is the largest!

Like the other giant planets, Jupiter is really a great ball of gases. It is composed mainly of hydrogen and helium. The planet has no solid surface beneath its atmosphere. The composition changes gradually from gas to liquid. Near the center of the planet, the pressure and temperature are so high that the liquid becomes solid, forming a small metallic core.

Jupiter appears as a very bright planet. The over-all color is yellow, but there are belts which vary from yellow to brown and there are tinges of blue and gray. Spots of color can be seen on the surface. One of the largest is known as the Great Red Spot. It is so large that the total area of Earth could fit on it! The Great Red Spot seems to be a permanent hurricane.

Jupiter has 63 known satellites. The four largest were discovered by Galileo in 1610 and are known as the Galilean moons. In 1979 *Voyager 1* photographed a bright, narrow ring with a fainter ring surrounding it.

Activities

Judge the choice of the name Jupiter for this planet.

Create a fact file about the Galilean moons. Include at least eight facts.

Jupiter is quite flattened. The diameter at the equator is 88,700 miles (142,800 kilometers). The distance from the north to south pole is only 83,000 miles (133,500 kilometers). Review the Planets Fact Sheet and guess why.

Jupiter is the most massive of all the planets, but it has very low density. Explain what this means.

A 150-pound person would weigh 380 pounds on Jupiter. Explain why.

© Educational Impressions, Inc.

Saturn

Like the other giant planets, Saturn has great size, low density, and a very extensive atmosphere. If viewed through a large telescope, Saturn appears as a yellow and gray banded body. Also, like Jupiter, Saturn bulges at the equator because of its rapid rotation on its axis.

Saturn's atmosphere, like that of Jupiter, is mostly hydrogen and helium; however, there is nothing like Jupiter's Great Red Spot, for Saturn's clouds are not as turbulent as those of Jupiter. The composition of the planet is also similar to Jupiter's. There is no hard surface. The gas gradually changes to liquid. Close to the center of the planet, the hydrogen becomes metallic. Scientists think there is a small iron core.

Saturn has at least 60 moons. Most of them are very small. In addition to moons, Saturn has a highly structured ring system, spanning 170,000 miles (275,000 kilometers). The rings are composed of billions of particles which orbit the planet. Some of the particles are no larger than a tiny grain; others are as large as a railroad car. Although the rings appear to be a few wide rings, they are really thousands of separate, narrow ringlets. For centuries people believed that Saturn was the only planet with rings. Since the *Voyager* space probes (1980), we know that all four of the giant outer planets have rings. The rings of Jupiter, Uranus, and Neptune, however, are not nearly as extensive as Saturn's ring system.

Activities

Titan is the second largest moon in the solar system. Find out what makes Titan unique.

Who was Saturn? Hypothesize as to why the ancients chose the name for this planet.

Explain what is meant by the Cassini Division.

If given a large enough ocean, Saturn would float! Why?

© **Educational Impressions, Inc.**

Uranus

Uranus, the seventh planet from the sun, was first observed in 1721 through a telescope. It appeared as a bluish-green disk without any unusual features. In 1869, dark absorption bands were discovered. It was learned that these bands were caused by methane in the atmosphere. Methane in the atmosphere is the major cause of the planet's blue-green color.

A very unusual feature of Uranus is the tilt of its axis. It is inclined at almost 98° from the plane of its orbit. In other words, the planet's poles lie almost in the plane of its orbit around the sun. It seems to rotate on its side!

The atmosphere of Uranus is mostly hydrogen and helium, with a larger percentage of helium than Jupiter or Saturn has. There is also a layer of methane ice clouds. Beneath the atmosphere, deep within the planet is a superheated water ocean, possibly about 6,000 miles (10,000 kilometers) deep. A core of molten rock was believed to exist, but many scientists now believe it is made of highly compressed liquid.

For many years, Uranus was thought to have only five moons and no rings. In 1977 rings were discovered, although it was not known how extensive they really are. *Voyager 2,* which passed over Uranus from November 1985 through February 1986, gave us a lot of new information about the moons and rings of the planet. We now know that in addition to the five major satellites, Uranus has at least twenty-two others. The ring system was found to comprise ten narrow rings of dark particles and one broad, diffuse ring. There are also 100 or more ringlets of dust-sized particles.

Activities

Find out the names of the major satellites of Uranus. What is unusual about their names?

Learn more about Voyager 1 and 2. Chart their journeys.

Uranus was discovered by William Herschel in 1721. He wanted to name the planet Georgium Sidus, or Georgian Star, after King George III of England. Many astronomers wanted to name it Herschel after its discoverer. Johann Bode proposed the name Uranus, which gradually became universally accepted. Which name (or another) would you have chosen? Why?

© Educational Impressions, Inc.

Neptune

Neptune is the last of the giant gas planets. In 1845 and 1846 two astronomers, Englishman John Couch Adams and Frenchman Urbain Leverrier, independently predicted the existence and location of an eighth planet. They saw that Uranus was being pulled from its true orbit and theorized that an unknown planet was the cause. Adams and Leverrier were right!

When viewed through a large telescope, Neptune appears as a small blue disk. *Voyager 2* photos showed many previously unknown features. A large dark storm system, named the Great Dark Spot, was seen in the southern hemisphere. By 1994 the storm system ceased to appear in the south, but a similar system appeared in the northern hemisphere. The probe also photographed cloud formations that appeared and vanished quickly, leading us to believe that, like Earth, Neptune has a dynamic, changeable weather system. Subsequent explorations showed that Neptune has at least thirteen satellites. Five rings were also revealed: two bright, narrow ones and three faint ones.

Neptune's atmosphere is mostly hydrogen and helium. Almost 3 percent of it, however, is methane. There are many cirrus clouds in the atmosphere. Scientists believe these clouds consist of methane crystals rather than water ice. These large amounts of methane are what give Neptune its deep blue color.

Activities

Explain how methane, a colorless gas, is responsible for the blue-green color of the atmosphere of Uranus and Neptune.

Evaluate the choice of the name Neptune for this planet.

Until the Voyager probes, scientists thought Neptune had only two satellites: Triton and Nereid. The orbit of Triton, the third largest moon in our solar system, is unique among the large satellites of our solar system. Explain.

Find out what is unusual about the orbit of Nereid.

© Educational Impressions, Inc.

Pluto, a Dwarf Planet!

For 76 years Pluto was called tthe ninth planet in our solar system! In 2006, however, it was reclassified as a dwarf planet. Pluto is so remote that astronomers still know little about it. Until fairly recently, most astronomers believed that Pluto was slightly larger than Mercury, which they thought was the smallest known planet in the solar system. With information obtained from the Hubble Space Telescope (HST), we learned that the planet probably has a diameter of about 1,416 miles (2,284 kilometers). As this is less than half of Mercury's diameter, Pluto is now known to be the smallest.

Pluto has a thin atmosphere of methane and nitrogen. The atmosphere varies in thickness according to its distance from the sun. When the planet is farthest from the sun, the atmosphere freezes. Images from the Hubble Space Telescope show polar ice caps. These ice caps are most likely frozen nitrogen. Beneath the atmosphere is a mantle of ice. Pluto seems to have a high density. This leads scientists to believe it has a large, rocky core.

Pluto has an unusual orbit. First of all, it is on its side, much like Uranus. Also, its average distance from the sun is 3.66 billion miles (5.9 billion kilometers), or 39.44 astronomical units (AU). Pluto's orbit is so elliptical that during 20 years of its 248.4-year orbit, Pluto is actually closer to the sun than is Neptune! This was the case from January 23, 1979, to March 15, 1999.

Pluto has one satellite, called Charon. It wasn't discovered until 1978. Charon revolves around the planet in the same time that it takes for Pluto to rotate once. For that reason, the same side of Pluto always faces its moon.

Activities

Find out the new definitions for *planet* and *dwarf planet*. Do you agree that Pluto belongs in the new classification?

Judge the choice of the names Pluto and Charon for the ninth planet and its moon.

Clyde Tombaugh, discoverer of Pluto, and other astronomers looked for a Planet X to explain the wanderings of the orbits of Uranus and Neptune. When Tombaugh discovered Pluto, they believed they had their explanation. When Pluto was found to be so much smaller than expected, some astronomers were led to believe that there is still a Planet X waiting to be discovered. Analyze their reasons.

© Educational Impressions, Inc.

Our Moon

The moon, Earth's only natural satellite, is about one-quarter the size of Earth. Its orbit is elliptical, just like those of the other bodies in our solar system. It takes 27 days, 7 hours, and 11.6 seconds to make one complete orbit.

The moon is our closest neighbor in space—about 240,000 miles (384,000 kilometers) away. Long ago, people had some strange ideas about the moon. Science-fiction stories were written about travel to the moon. Eventually, unmanned spacecraft were sent; we learned a lot from the pictures they sent back. Then—on July 20, 1969—science-fiction became a reality. As part of the Apollo 11 mission, astronauts Neil Armstrong and Edwin Aldrin, Jr., landed on the moon! Third crew member Michael Collins remained in the command module.

Armstrong, the first human to set foot on the moon, described its surface as fine and powdery. He and Aldrin set up machines to send back information to Earth. They also gathered rocks and moon dust for scientific study.

From Apollo 11 and other missions, we've learned a lot about the moon. It's a gray and lifeless place—a place of high, sharp mountains and broad, flat plains. Some of these plains are called maria, or seas, although there is no water in them. It's also a place of long, narrow, trench-like valleys, called rilles. Many craters dot the surface. Some of the craters may be the result of volcanoes whose tops were blown off or collapsed. Others are probably the result of asteroids, meteors, or comets which hit the moon's surface.

From the rock samples brought back by astronauts we know that moon rocks have a slightly different composition from those on Earth. Earth's crust and the moon's crust were learned to have similar amounts of oxygen, silicon, and aluminum. The moon's crust, however, was shown to have more iron and titanium, but less alkali metals, carbon, and nitrogen. No water was found on the moon. In fact, the only hydrogen found was carried in by the solar wind.

The atmosphere of the moon is very thin; therefore, the sky is always black. There is no oxygen in the atmosphere. That's one of the reasons astronauts have to wear space suits while on the moon. Another reason is to protect themselves from the extremely hot daytime temperatures and the extremely cold nighttime temperatures. They also have to carry devices to communicate with each other, for there is no air to carry sound.

© **Educational Impressions, Inc.**

Activities

Find out what is meant by a synodic month and a sidereal month. Why are they different?

In order to fly astronauts to the moon, a powerful rocket was needed. Find out the name of the rocket that launched Apollo 11.

We always see the same side of the moon here on Earth. Explain why.

Explain why astronauts seem to bounce as they walk on the moon.

Create a lesson to demonstrate to a younger child why the moon and the sun appear to be about the same size to us.

Name our moon. Justify your choice.

Daytime and nighttime each lasts about two weeks on the moon. Daytime temperatures reach over 212°F (100°C). Nighttime temperatures drop to about −240°F (−150°C). Explain why.

During an eclipse of the moon, the bright moon gradually darkens in the night sky. Draw a diagram to explain why this occurs.

As the moon orbits Earth, its shape seems to change. These changes are called phases. The phases are caused because different areas of the side of the moon which faces Earth are lighted by the sun. Create a chart that shows the phases of the moon.

© Educational Impressions, Inc.

A One-sided Point of View

Do the following experiment:

1. Make a fist and pretend it is Earth.

2. Take a tennis ball or similarly sized ball and pretend it is the moon.

3. Put an X on the side of the ball (moon) that faces your fist (Earth).

4. Slowly rotate the ball (moon) on its axis.

5. At the same time and at the same speed, move the ball (moon) around your fist (Earth)—just like the moon revolves around Earth.

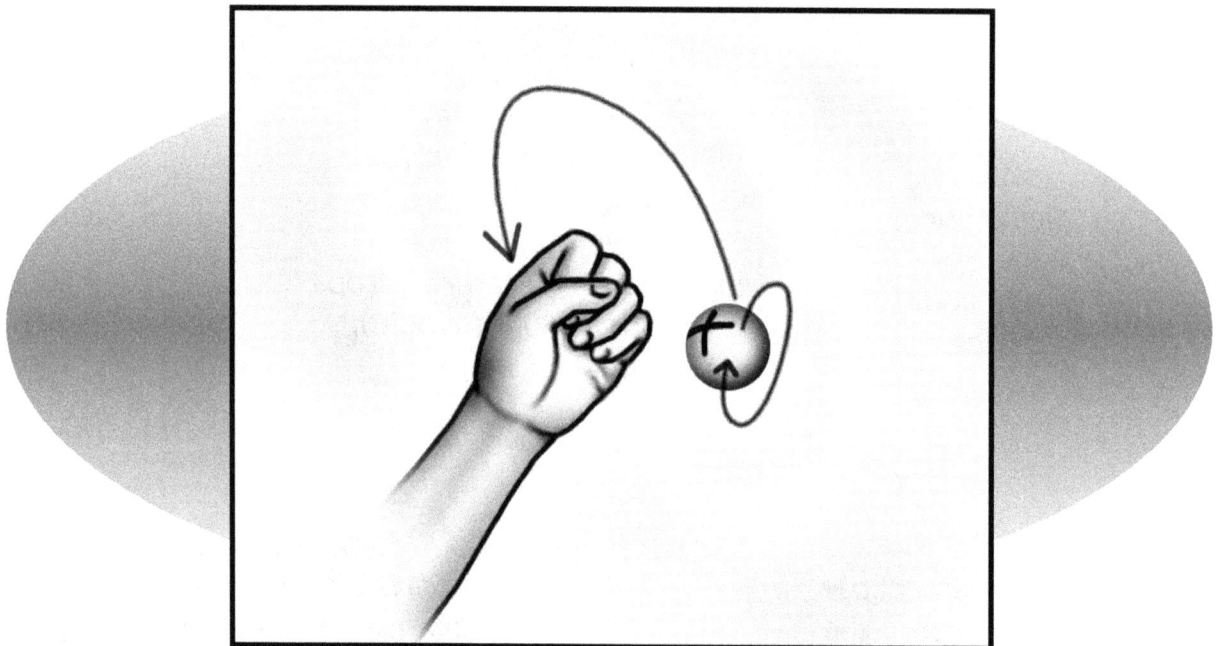

Describe the results of your experiment. What conclusion can you reach?

_____

_____

_____

_____

© **Educational Impressions, Inc.**

Tides

Although the moon's gravitational pull is less than Earth's, it still affects Earth in an important way. It causes the waters of the oceans to rise and fall. These changes are called tides.

The oceans have two high tides and two low tides each day. When the moon is directly above the ocean, the tide is highest; the moon pulls the ocean away from Earth, and the ocean bulges toward the moon. At the same time, the ocean on the opposite side of Earth also has high tide; in this case, Earth is pulled away from the ocean. The water is drawn from other parts of the oceans; those parts experience low tide.

Draw a diagram to illustrate how the moon affects Earth's oceans.

Historic Words

Neil Armstrong, the first human being to set foot on the moon, sent a message to those of us who remained on Earth. Decipher the puzzle and find out what this message was. (You may go in any direction—up, down, sideways, or diagonally—but do not skip over any letter.)

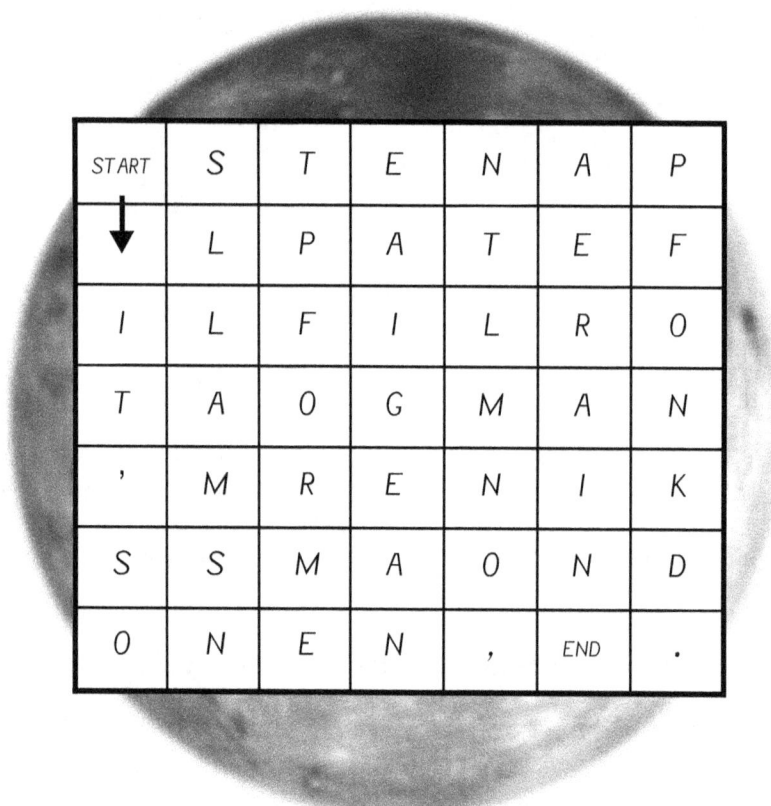

START	S	T	E	N	A	P
↓	L	P	A	T	E	F
I	L	F	I	L	R	O
T	A	O	G	M	A	N
'	M	R	E	N	I	K
S	S	M	A	O	N	D
O	N	E	N	,	END	.

" ___ ___ ' ___ ___   ___ ___ ___ ___

___ ___ ___ ___ ___

___ ___ ___ ___ ___ ___ ___ , ___ ___ ___

___ ___ ___ ___   ___ ___ ___

___ ___ ___ ___ ___ . "

© Educational Impressions, Inc.

Asteroids

In addition to the planets and their moons, there are other bodies that revolve around the sun and are, therefore, part of our solar system. Asteroids are small, irregularly shaped, solid objects. Because they orbit the planet directly, they are sometimes called minor planets. Most of the asteroids are found in a swarm between Mars and Jupiter. This swarm is called the asteroid belt. Asteroids range in size. Some are tiny. Others are hundreds of miles (kilometers) in diameter. Ceres, the largest known asteroid, is about 488 miles (785 kilometers) long. It was the first asteroid to be discovered. Today we know that there are over 100,000 asteroids, more than 20,000 of which have been identified. When an asteroid's orbit is established, it is assigned a number. More than 5,000 have been numbered. Some, like Ceres, are also named.

Activities

When Pluto's small size was learned, some scientists argued that it should be reclassified as an asteroid. (At that time the classification Dwarf Planet did not exist. Choose a side and present an argument.

Many scientists believe that the asteroids were too close to Jupiter to grow together as a planet or larger satellite. Explain.

Icarus, a large asteroid, comes relatively close to the sun. At times it is less than 18,600,000 miles (30,000,000 kilometers) away. Research Icarus to find out why the asteroid was so named.

© Educational Impressions, Inc.

Comets

Comets are other celestial bodies that are part of our solar system. We can see comets only when they are in the part of their orbit that takes them relatively close to the sun.

The nucleus, or center part of the comet's head, is composed mostly of ice and rocks. As the comet approaches the sun and some of the ice turns to gas, a hazy cloud of gas and dusty particles, called a coma, forms around the nucleus. The closer the comet gets to the sun, the greater the production of gas.

Many comets develop tails. Some comet tails are thousands—even millions—of miles long! Part of the tail usually appears yellowish; it is the dust tail. The other part is bluish; it is the gas tail. The gas tail probably consists of ammonia, methane, and carbon dioxide. The tail of a comet is the most noticeable feature to us on Earth. Some are quite spectacular!

The time it takes for a comet to orbit the sun varies greatly. Some comets, called short-period comets, take up to a few decades. For example, Comet Encke, which has the shortest orbit, takes slightly over three years. Others, called long-term comets, take many thousands of years.

Scientists aren't certain where comets originate. Many believe that there is a belt of comets surrounding the solar system. The gravitational effect of passing stars might cause them to come out of the cloud towards the sun. Once within the orbits of the planets, they could be held in shorter orbits around the sun.

Comets are named after the people who discover them. One of the most famous comets is Halley's Comet, named for Edmond Halley. Halley noticed that the orbits of the comets observed in 1531, 1607, and 1682 were almost identical. He concluded that it was the same comet viewed each time. Halley correctly predicted its return in 1758.

Activities

Guess why a comet's tail always faces away from the sun.

The first recorded sighting of what was later named Halley's Comet dates from 240 B.C. Create a conversation between two ancients viewing the phenomenon.

When comets are far from the sun, they are not the bright, glowing objects we usually visualize. In fact, astronomers sometimes refer to them as "dirty snowballs." Create three similes or three metaphors to describe comets or other bodies in our solar system.

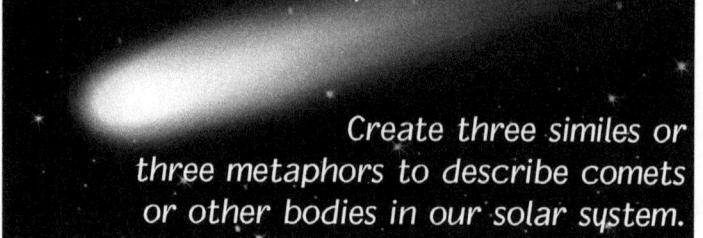

© Educational Impressions, Inc.

Meteoroids, Meteors, and Meteorites

Meteoroids are bits and pieces of comets or asteroids before they enter Earth's atmosphere.

Meteoroids become **meteors** when they enter Earth's atmosphere, falling at speeds of 6 to 45 miles (10 to 72 kilometers) per second. Because they fall so quickly, they get so hot that they flare up as streaks of light and then evaporate.

A group of meteors that appear together in a swarm and seem to have a common origin is called a **meteor shower.** Most meteor showers are believed to be the debris of comets.

Those meteors larger than a small stone appear especially bright and are called **fireballs.** Sometimes fireballs explode in the sky.

Meteors that are large enough to travel through Earth's atmosphere and hit the ground are called **meteorites.** There are three kinds of meteorites: iron, stone (silicate), and more rarely, stony irons. Meteorites are probably remnants of asteroids.

Activities

Draw a picture that illustrates a meteor, a meteor shower, and a meteorite.

Sometimes meteors are called "falling stars" or "shooting stars."
Judge the use of these terms.

© Educational Impressions, Inc.

The Milky Way

A galaxy is a vast collection of stars. There are billions of galaxies in the universe. Our sun is in the galaxy we call the Milky Way. It is a spiral galaxy, which means it has arms resembling a pinwheel. Like other spiral galaxies, the Milky Way is very bright. Its nucleus (center) contains many stars.

Activities

Research the Milky Way. Make a 5-card fact file about the galaxy.

The ancient Greeks named the galaxy the Milky Way. In their mythology, it was said to be milk spilled by the goddess Hera as she nursed her infant Heracles. Create an original legend to explain the existence of our galaxy or solar system.

Most stars are more than 65 light-years away. The nearest star, however, is 3.3 light-years away. Find out the name of that star.

Define light-year.

© **Educational Impressions, Inc.**

Exploring Space

Use the clues to help you unscramble the words and answer the question.

1. What is the name of the first artificial satellite?   ANSWER: **P S T U N I K**

    CLUE #1: It was launched by the Soviet Union.
    CLUE #2: It was launched on October 4, 1957.
    CLUE #3: In Russian it means "fellow traveler."

    ____ ____ ____ ____ ____ ____ ____

2. Who discovered the four large moons that orbit Jupiter?   ANSWER: **A G L I E L O**

    CLUE #1: He was an Italian scientist.
    CLUE #2: He lived from 1564–1642.
    CLUE #3: He built a refracting telescope.

    ____ ____ ____ ____ ____ ____ ____

3. What is the name of the largest space telescope put into orbit?   ANSWER: **U H B B E L**

    CLUE #1: It was launched in 1990.
    CLUE #2: It is basically a large video camera.
    CLUE #3: There was a problem with the mirrors.

    ____ ____ ____ ____ ____ ____

4. What space probe is scheduled to reach Saturn in 2004?   ANSWER: **S A I C S N I**

    CLUE #1: It was launched in 1997.
    CLUE #2: It is also expected to study Saturn's largest moon, Titan.
    CLUE #3: It was named for the 27,000-mile (3,000 km) gap in Saturn's rings.

    ____ ____ ____ ____ ____ ____ ____

5. What space probe is designed to study the sun?   ANSWER: **E L U Y S S S**

    CLUE #1: It was launched in 1995.
    CLUE #2: It was named for a king in Roman mythology; he reached home
                  after ten years of wandering.
    CLUE #3: In Greek the character's name was Odysseus.

    ____ ____ ____ ____ ____ ____ ____

Planetary Syllogisms

A syllogism is a type of deductive argument. It is based on a major premise (A), a minor premise (B), and a conclusion (C). In a valid argument, the conclusion must be in agreement with and based upon the previous statements.

Decide whether each of the following arguments is valid or invalid and write the appropriate word. If invalid, explain why.

1. A. Planets do not emit self-generated light; they are illuminated by the star around which they revolve.
    B. Venus is a planet.
    C. Therefore, Venus does not emit self-generated light.

    _____

2. A. Some planets have one or more satellites, or moons, which orbit them.
    B. Mercury is a planet.
    C. Therefore, Mercury has at least one moon.

    _____

3. A. Jupiter is the largest planet in our solar system.
    B. Saturn is a planet in our solar system
    C. Therefore, Jupiter is larger than Saturn.

    _____

4. A. All of the planets in our solar system except Earth were named for mythological characters in ancient Greek or Roman mythology.
    B. Io is a character in the mythology of ancient Greece.
    C. Therefore, Io is a planet in our solar system.

    _____

                                    © **Educational Impressions, Inc.**

Complete these syllogisms with valid conclusions.

5. A.   The closer a planet is to the sun, the less time it takes for the planet to revolve around the sun.

    B.   Earth is closer to the sun than is Mars.

    C.   Therefore, _____

    _____

6. A.   The atmosphere of the giant outer planets are mostly hydrogen and helium.

    B.   Neptune is a giant outer planet.

    C.   Therefore, _____

    _____

7. A.   The farther a planet is from the sun, the longer it takes for the planet to orbit the sun.

    B.   It takes Jupiter over 29 Earth-years to orbit the sun; it takes Uranus almost 164 Earth-years to orbit the sun.

    C.   Therefore, _____

    _____

Now create three syllogisms about the solar system. Exchange with classmates to solve.

1. A.   _____

    B.   _____

    C.   _____

2. A.   _____

    B.   _____

    C.   _____

3. A.   _____

    B.   _____

    C.   _____

© Educational Impressions, Inc.

Create Math Problems

Create a math problem for each set of data. You do not have to use all the information, and you may add information as necessary. Be sure to provide a detailed solution for each problem.

PROBLEM NO. 1

Mercury's diameter is about 3,030 miles.

Earth's diameter is about 7,900 miles.

Jupiter's diameter is about 86,000 miles.

The Sun's diameter is about 860,000 miles.

The Moon's diameter is about 2,040 miles.

PROBLEM NO. 2

Mercury rotates in about 58.6 Earth-days.

Venus rotates in 243 Earth-days.

Earth rotates in 1 Earth-day or 24 Earth-hours.

Mars rotates in 24.6 Earth-hours.

Jupiter rotates in 9.9 Earth-hours.

Saturn rotates in 10.5 Earth-hours.

© **Educational Impressions, Inc.**

What's the Question?

What's the Question? is similar to the TV show "Jeopardy" in that the information is given in the form of the statement and the student responses are in the form of questions. The questions in Part 1 are worth 5 points each, and those in Part II are worth 10 points each.

Divide the class into teams of 4 or 5 students. The teacher may act as leader, or you may want to choose a student leader. The leader asks the first group a question from Part I. Whoever raises his or her hand first gets to answer. If the student answers correctly, 5 points are added to the team total. If the student answers incorrectly, 5 points are deducted. If no one wants to answer, the leader gives the correct answer and the totals remain the same. If a team does not give a correct answer, the same question is then asked to the next group. If no group gets it right, the leader gives the correct answer.

When all the questions from Part I have been completed, the same rules are followed for Part II.

ANSWERS TO "SOLAR SYSTEM WHAT'S THE QUESTION?"

PART I: 5 points each

1. What is a star?
2. What is "emit self-generated light"?
3. What is the sun?
4. What is the Milky Way?
5. What is Earth?
6. What is Jupiter?
7. What is Mercury?
8. What is Neptune?
9. What is Saturn?
10. What is Mercury or Venus?
11. What are tides?
12. What are phases?
13. What are craters?
14. What are maria?
15. What happens when the moon passes into Earth's shadow?
16. What are Mars and Jupiter?
17. What is Ceres?
18. What is a streak of light?
19. What is a meteorite?
20. What are comets?
21. What is Earth?
22. Who was Copernicus?
23. Who was Isaac Newton?
24. Who was Kepler?
25. What is a telescope?

PART II: 10 points each

1. What is a dwarf star?
2. What are giant stars?
3. What is luminosity?
4. What is the Milky Way?
5. What is the corona?
6. What is Ganymede?
7. What is Pluto?
8. What is black?
9. What is Saturn?
10. What is Titan?
11. What is the nucleus?
12. What is "after the person who discovered it"?
13. What is Halley's?
14. What is the bright part of the comet?
15. What is solar wind?
16. Who is Neil Armstrong?
17. What is Mars?
18. What are *Voyager 1 and 2*?
19. What is a meteorite?
20. What is *Saturn V*?
21. What is a solar eclipse?
22. What is 1930?
23. What is Neptune?
24. What is the universe?
25. What is National Aeronautics and Space Administration?

What's the Question?

Part I: 5 points

Our Sun	Planets	Our Moon	Asteroids & Meteoroids	History of Astronomy
1. The sun is just an average one.	6. It is the largest planet in our solar system.	11. Our moon affects Earth's oceans by causing them.	16. The 100-mile-wide band known as the asteroid belt is located between these two planets.	21. Ancients believed this was the center of our universe.
2. A star does this, but a planet does not.	7. It is the closest planet to the sun.	12. The different areas of our moon lit by sunlight are called this.	17. It's the largest known asteroid.	22. This Polish astronomer published a sun-centered theory of the universe.
3. It is the largest body in our solar system.	8. This planet is sometimes farther from the sun than Pluto is.	13. Most of them were caused when smaller bodies crashed into the moon's surface.	18. A meteor appears as this when it burns up in the atmosphere.	23. This English scientist's theory of gravity explained why planets orbit the sun as they do.
4. Our sun is in this galaxy.	9. At one time this was the only planet in the solar system known to have a ring system.	14. In spite of their name, these lowland plains have no water.	19. We call a large meteoroid this when it strikes Earth.	24. His laws of planetary motion explained that planets have elliptical orbits.
5. The sun is 93 million miles from this planet.	10. It's one of two planets in our solar system with no moon.	15. A lunar eclipse occurs when this happens.	20. Meteor showers are thought to be the debris of these celestial bodies.	25. The type built by Galileo was a refracting one.

© Educational Impressions, Inc.

What's the Question?

Our Sun	Planets	Comets	Space Exploration	Pot Luck
1. Our sun is classified as this kind of star.	6. Jupiter's largest moon, it is also the largest moon in our solar system.	11. It is the solid part of a comet.	16. He was the first human to set foot on the moon.	21. It occurs when the moon passes between Earth and the sun .
2. These stars are more massive than our sun and also give off more light.	7. Once called the ninth planet, in 2006 it was reclassified as a dwarf planet.	12. It's how a comet is named.	17. The *Viking* missions in 1976 gave us a lot of information about this red planet.	22. It's the year Pluto was discovered.
3. It is the term for the quality of emitting self-generating light.	8. Because the moon's atmosphere is so thin, the sky always appears this color.	13. This famous comet returns every 76 years.	18. These space probes, launched in 1977 and 1979, flew over all the outer planets except Pluto.	23. This planet was named for the Roman god of the seas because of its deep blue color.
4. It is the outermost atmosphere of the sun and appears as a faint halo.	9. This planet has the most known satellites.	14. It is the brightest part of a comet.	19. This rocket launched *Apollo 11* to the moon.	24. It's what we call all of space and everything in it.
5. This term refers to the relatively dark spots that appear in groups on the surface of the sun.	10. Saturn's largest moon, it is the second largest satellite in the solar system.	15. A comet's tail always faces away from the sun because of this.	20. The Age of Space Exploration began with the launch of this Soviet satellite in 1957.	25. NASA is an acronym for this agency.

Solar System Crossword Puzzle

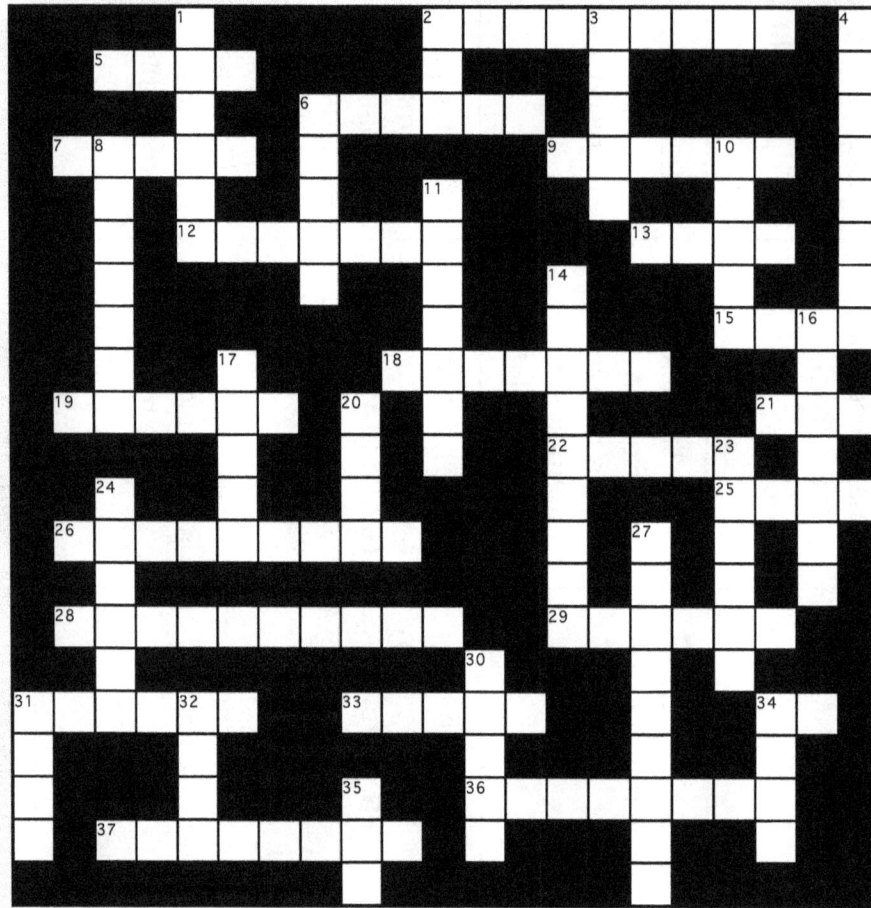

ACROSS

2. A relatively smaller body orbiting a planet.
5. Our sun is one.
6. A non-luminous body orbiting a star.
7. Second planet from the sun.
9. The recurring apparent forms of the moon.
12. Named for the Roman god of the seas.
13. Sometimes called the Red Planet.
15. Compared to Mercury, Jupiter is this.
18. The largest planet in our solar system.
19. It appears as a streak of light.
21. It is ___time on the half of a planet facing the sun.
22. Saturn was once the only planet known to have them.
25. What a planet rotates on.
26. Last name of first human to set foot on moon.
28. Venus has a thick one, the moon a thin one.
29. The sixth planet from the sun.
31. Faintly colored luminous ring around the sun.
33. Path of body as it revolves around another.
34. Mercury and Venus have ___ moon.
36. This gas is most abundant in the atmosphere of the giant outer planets.
37. The galaxy in which our solar system is located.

DOWN

1. Titan orbits this planet.
2. The center of our solar system.
3. A planet does not emit this.
4. All existing things, including all of the galaxies.
6. Now called a dwarf planet.
8. The partial or complete obscuring of one celestial body by another.
10. Third planet from the sun.
11. Second smallest planet.
14. Sometimes called minor planets; most are between Mars and Jupiter.
16. Force which attracts celestial bodies toward each other.
17. A tail forms from its coma as it nears the sun.
20. It causes the tides on Earth.
23. The Ringed Planet.
24. A pit or hole in the ground created by a meteorite.
27. The scientific study of the universe.
30. It is ___time on the half of a planet facing away from the sun.
31. The central portion, such as that of a planet.
32. First name of 26 Across.
34. Number of planets in our solar system.
35. The giant outer planets are mostly in this state.

© **Educational Impressions, Inc.**

Answers and Background Information

Our Solar System: Ancient Views (Page 8)
In **Ptolemy's view,** Earth was at the center of the universe. The moon and planets (Saturn was the farthest known planet at that time.) revolved around Earth in circular orbits. To explain why the starry background changed nightly, he theorized that the moon and the planets also revolved in small circles. Like others of his time, he believed the circle was the only form perfect enough for the celestial bodies.

Copernicus believed the sun (Helios) was the center of the universe. His theory implied the enormous size of the universe. Although he was correct in believing that the moon and the planets revolved around the sun, he was wrong in believing that the stars, too, revolved around the sun. Also, he still wrongly believed that the orbits were circular. (Saturn was still the farthest known planet.)

Ptolemy held a geocentric view. **Geocentric** means "earth-centered." Copernicus held a heliocentric view. **Heliocentric** means "sun-centered."

Astronomical Match-up (Page 9): ASTRONOMY

The Sun: Just an Average Star! (Page 12)
A **solar eclipse** of the sun takes place when there is a new moon and when the moon is between Earth and the sun. Once in a while the shadow of the moon passes over part of Earth and there is a partial eclipse. Even more rarely, the entire shadow reaches Earth, and there is a total eclipse of the sun.

Photosynthesis is the process by which green plants synthesize carbohydrates from carbon dioxide and water using light as an energy source. Plants, therefore, depend on the sun's light for growth. If there were no plants, then the animals that eat them could not exist. Without those animals, animals that eat other animals would not exist either.

An **astronomical unit** is a unit of length used in measuring astronomical distances. One AU is equal to the mean distance of Earth to the sun, about 93 million miles.

Sunspots are cooler areas on the surface of the sun. They appear as cool, dark patches. Every 11 years the number of sunspots is greatest. **Solar flares** are violent explosions occurring over sunspots; they last only a few minutes, but they do affect Earth's magnetic field. **Faculae** are clouds of glowing hydrogen; they lie just above the surface and appear around the sunspots.

About the Planets (Page 15)
Daytime and nighttime are caused because the planets rotate on their axis as they revolve around the sun. It is daytime on the half of the planet facing the sun. It is night on the half turned away from it.

Planets and **stars** are celestial bodies. Both are spherical. Stars are larger than the planets that revolve around them. Stars generate energy by nuclear reactions in their interiors. Planets do not. Only stars emit self-generated light. Planets revolve around a star and are held in their orbits by the star's gravity.

Mercury (Page 16)
Mercury is a small, rocky planet; it is covered with craters. Like other terrestrial planets, it has a crust, mantle, and hard metal core. The atmosphere on Mercury is thin.

The temperature drop on Mercury can be explained because the planet has a very thin atmosphere. There is not enough of an atmosphere to hold in the heat.

The Caloris Basin is a 800-mile-wide crater which formed when a 100-mile-wide rock collided with Mercury. Lava from the mantle spilled out and flattened the crater surface.

Mercury was named after the Roman messenger of the gods. The planet was likely so named because of its speedy travel around Earth.

Venus (Page 17)
Venus is closer to the sun. Both are inner, terrestrial planets. Venus has a denser atmosphere, which is mostly carbon dioxide. Earth's atmosphere contains oxygen. Temperatures are higher on Venus. Venus is tilted on its axis at a much greater angle, appearing to rotate backwards. Earth is larger. Venus has no moons; Earth one. Venus orbits the sun more quickly but takes much longer to rotate on its axis.

The greenhouse effect causes the high surface temperatures on Venus in the following way: Venus has a thick carbon dioxide atmosphere. The clouds allow some of the sunlight to filter through. As the surface heats up, the energy is radiated away as infrared radiation, which is absorbed by the lower atmosphere. Just as glass traps heat energy in a greenhouse, the clouds trap the heat energy in the atmosphere.

Plate tectonics is based on the theory that Earth's surface comprises a small number of large, semirigid sections that float across the mantle. Seismic activity and volcanism occur mainly where the sections meet.

Venus is unique because it is the **only planet whose rotation on its axis takes longer than its revolution** around the sun.

On Venus the **sun rises in the west and sets in the east** because of the extreme tilt of the axis.

Venus was named for a goddess of ancient Rome. She was associated with Aphrodite, the beautiful Greek goddess of love. It was probably given the name because the planet appears as a beautiful, bright object in the sky.

Earth (Page 18)

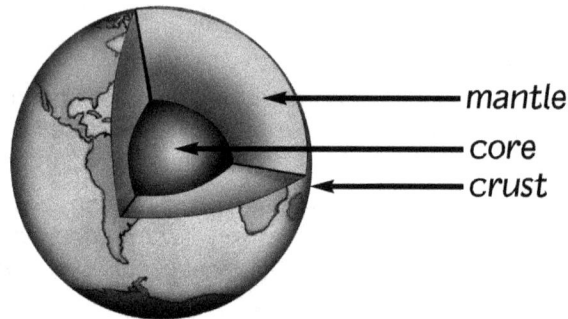

mantle
core
crust

Two hundred million years ago all the continents were joined into one continent, called **Pangaea**.

Earth is unique in the solar system because it supports life and has liquid water on its surface.

Mars (Page 20)
In Greek mythology **Phobos** ("fear") and **Deimos** ("terror") were the sons of Ares, the Greek equivalent of Mars, the ancient Roman god of war.

The movement of plates tends to relocate centers of mountain-building activity.

Jupiter (Page 21)
Jupiter was named for the king of the gods in Roman mythology. Jupiter is the largest planet in our solar system.

Facts about Jupiter's **Galilean moons** will vary, but may include the following: They are the largest of Jupiter's moons. The four Galillean moons in order of their distance from Jupiter are Io, Europa, Ganymede, and Callisto. In size order they are Ganymede, Callisto, Io, and Europa. Ganymede is the largest moon in our solar system. Ganymede has an icy surface, which is covered by craters. Callisto is the darkest and iciest moon. Io has sulfur-rich volcanoes. Europa has a layer of ice which makes its surface the smoothest of any moon or planet in our solar system. The Galilean moons were discovered by Galileo in 1610.

The amount of material in a particular space is called is **mass. Density** is how heavy something is relative to its size. The closer together the atoms are, the denser the material.

The stronger the pull of gravity, the more an object (or person) weighs. Because the strength of Jupiter's gravity is stronger than Earth's, one would **weigh more on Jupiter than on Earth.**

Saturn (Page 22)
Titan is the only moon in the solar system with a substantial atmosphere.

Saturn was named for the ancient Roman god of agriculture, who was identified with the Greek god Cronus. Cronus was the father of Zeus, who was the Roman equivalent of Jupiter. Zeus dethroned Saturn, and Saturn vanished from Earth.

© **Educational Impressions, Inc.**

The Cassini division is the name of the 2,000-mile (3,000-kilometer) gap that separates the dimmer, inner part of Saturn's ring system from its bright, outer part. It was named for the Italian-born French astronomer Jean Dominique Cassini, who first discovered the gap in 1675.

Saturn would float because its gases are not tightly compressed, causing its density to be less than that of water.

Uranus (Page 23)
Uranus 's moons were named for Shakespearean characters: Miranda, Ariel, Umbrial, Titania, and Oberon. Most of the larage satellites in our solar system were named after mythological characters.

Uranus was named for the mythological father of Saturn. In Greek mythology, Uranus was the earliest supreme god, representing the sky or heaven. It was probably a good choice because the other planets in our solar system were also named for figures in Greek and Roman mythology.

Voyager 1 was launched from Kennedy Space Center on September 5, 1977. On March 5, 1979, it made its closest approach to Jupiter. On November 12, 1980, it made its closest approach to Saturn. It headed above the plane of our solar system toward interstellar space. **Voyager 2** was launched from Cape Canaveral on August 20, 1977. On July 9, 1979, it made its closest approach to Jupiter. On August 25, 1981, it made it closest approach to Saturn. On January 24, 1986, it made its closest approach to Uranus. On August 25, 1989, it made its closest approach to Neptune. It headed below the plane of our solar system toward interstellar space. NOTE: Scientists hope that at least one of them will reach the heliopause, the outer limit of sun's magnetic influence.

Neptune (Page 24)
The **methane,** although itself colorless, absorbs the red light from the white sunlight, leaving a blue-green color in the reflected sunlight.

Neptune was named for the ancient Roman god of the sea because of its deep blue color.

Triton orbits in a retrograde, or backward, direction around Neptune. The normal rotation of bodies in the solar system is west to east. It rotates east to west.

Nereid has the most elongated orbit of any moon. It ranges from .9 to 6 million miles (1.4 million and 9.7 million kilometers).

Pluto (Page 25)
The 2006 definition of "planet" by the International Astronomical Union states that a planet is a celestial body that is in orbit around the Sun; has sufficient mass so that it assumes a hydro-static equilibrium (a nearly round) shape; and has " cleared the neighborhood " around its orbit. A non- satellite body fulfilling just the first two criteria is classified as a " dwarf planet."One fulfilling only the first criterion is called a "small solar system body "

Pluto was the ancient Roman god of the underworld. He was associated with Hades in Greek mythology. Charon was the ferryman who conveyed the dead to Hades over the river Styx.

Some scientists believe there is still a **Planet X** because they think Pluto's gravity is too small to affect the orbits of worlds other than its own moon.

Our Moon (Page 27)
The **synodic month** is the length of time it takes for the moon to return to the same phase: 29 days, 12 hours, 44 minutes, and 2.8 seconds. The **sidereal month** is the period in which the moon completes an orbit and returns to the same position in the sky: 27 days, 7 hours, 43 minutes, and 11.6 seconds. They differ because Earth moves in its orbit around the sun in the same direction as the moon.

We see the same side of the moon because the moon rotates on its axis in about the same time it takes for it to complete one orbit around Earth.

Saturn V, a 3-stage rocket, launched *Apollo 11.*

The lesson should demonstrate that the **farther away something is, the smaller it looks.**

Astronauts seem to bounce because the force of gravity on the moon is much less than on Earth.

There is no atmosphere to filter the sun's rays; therefore, **daytime temperatures soar.** Neither is there an atmosphere to hold in the heat; therefore, the heat quickly escapes from the side facing away from the sun and **nighttime temperatures drop.**

The phases of the moon:

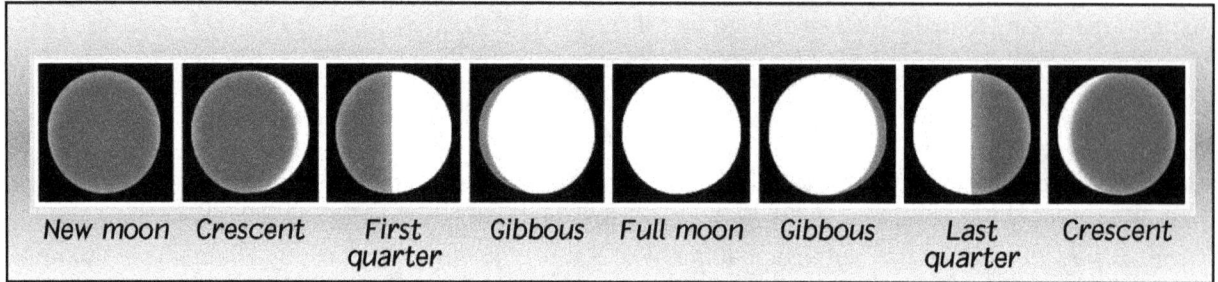

| New moon | Crescent | First quarter | Gibbous | Full moon | Gibbous | Last quarter | Crescent |

Once in a while, when there is a full moon and Earth is between the sun and the moon, the moon enters Earth's shadow and an **eclipse of the moon** occurs. The bright moon gradually darkens in the night sky.

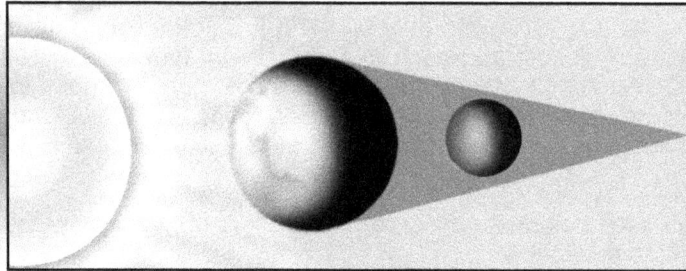

A One-sided Point of View (Page 28)
The experiment should show that we always see the same side of the moon because the moon rotates on its axis in about the same time it takes for it to complete one orbit around Earth.

Tides (Page 29)

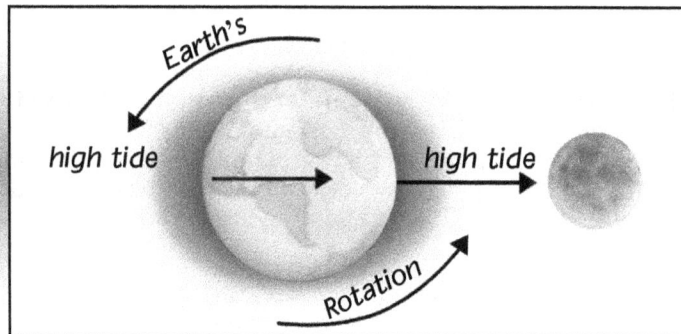

Historic Words (Page 30)
Neil Armstrong said, "It's one small step for man, one giant leap for mankind."

Asteroids (Page 31)
For calling Pluto an asteroid: It is smaller than other planets. Its gravity doesn't seem to affect other planets. Its orbit is so elliptical that it actually gets closer to the sun than Neptune for a period of time during each revolution. **For continuing to call Pluto a major planet:** It is spherical. Mercury and Mars are also relatively small. It is part of our accepted knowledge.

The **asteroids didn't grow together as major planets** because Jupiter's overpowering pull would have prevented them from growing into major planets.

In Greek mythology, **Icarus** was the son of Daedalus. Daedalus made artificial wings for him. He wore the wings in escaping from Crete, but flew too close to the sun. The wax which was used to fasten the wings melted, and Icarus fell into the Aegean Sea.

Comets (Page 32)
The **solar wind** keeps the comet's **tail pointed away** from the sun.

                    © **Educational Impressions, Inc.**

Meteoroids, Meteors, and Meteorites (Page 33)

meteor

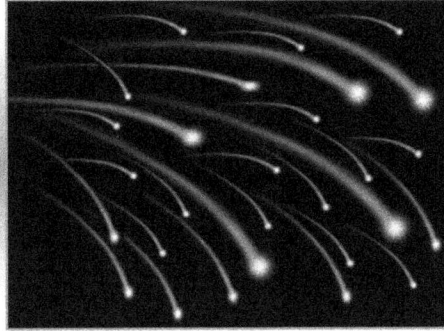
meteor shower

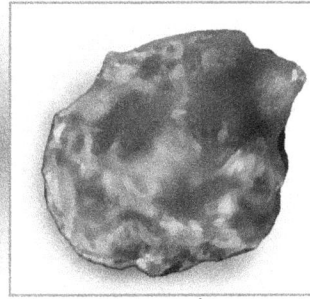
meteorite

"**Falling stars**" and "**shooting stars**" are misnomers, for meteors have nothing to do with stars. Before astronomers knew what they were, people thought they were stars falling from the sky.

The Milky Way (Page 34)
Facts will vary but may include the following: Although often used to refer to the entire galaxy, "The Milky Way" really refers to the portion visible to the naked eye. The Milky Way is seen as a broad band of faint light in the night sky. The visible disk is about 100,000 light-years wide, but it is only about 2,000 light-years thick. The galaxy has a spiral structure. It is estimated to contain at least 200 billion stars. It is one of the largest spirals. Our sun takes about 225,000 million years to orbit the galaxy's center. The sun is a little more than 30,000 light years from the center of the galaxy.

Proxima Centaur, our nearest star, is 3.3 light years from Earth.

A light-year is the distance light travels in a vacuum in one year, about 5.88 trillion miles (9.46 trillion kilometers).

Exploring Space (Page 35)
1. SPUTNIK   2. GALILEO   3. HUBBLE   4. CASSINI   5. ULYSSES

Planetary Syllogisms (Page 36–37)
1. Valid   2. Invalid (The major premise said "some," not "all.")   3. Valid
4. Invalid (It doesn't say all mythological character names are also planets.)
5. Therefore, it takes less time for Earth to orbit the sun than it does for Mars to orbit the sun.
6. Therefore, Neptune's atmosphere is mostly hydrogen and helium.
7. Therefore, Neptune is farther from the sun than Saturn.

What's the Question (Pages 39–41)
The answers are given on page 39.

Solar System Crossword Puzzle (Page 42)

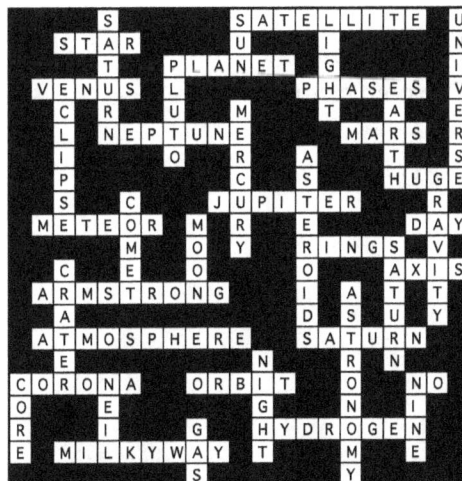

Bibliography

Branley, Franklyn M. *Neptune: Voyager's Final Target.* New York: HarperCollins Publishers, 1992.

_____. *Uranus: The Seventh Planet.* New York: Thomas Y. Crowell Junior Books, 1988.

Ford, Henry. *The Young Astronomer.* New York: DK Publisher, Inc., 1998.

Fradin, Dennis Brindell. *The Planet Hunters.* New York: Simon & Schuster, 1997.

Harris, Alan and Paul Weissman. *The Great Voyager Adventure.* Englewood Cliffs, NJ: Julian Messner, 1990.

Muirden, James. *Stars and Planets.* New York: Kingfisher Books, 1993.

Parker, Steve. *Galileo and the Universe.* New York: HarperCollins Publishers, 1992.

Ridpath, Ian. *The Concise Handbook of Astronomy.* New York: W.H. Smith Publishers, Inc., 1986.

Riley, Peter and Lawrence T. Lorimer. *Our Solar System.* Westport, CT: Joshua Morris Publishing, Inc., 1998.

Scott, Elaine. *Close Encounters: Exploring the Universe with the Hubble Space Telescope.* New York: Hyperion Books for Children, 1997.

Simon, Seymour. *Destination, Jupiter.* New York: William Morrow and Company, Inc., 1998.

_____. *The Sun.* New York: William Morrow and Company, Inc., 1986.

http://www.bbc.co.uk/science/space/solarsystem/
http://science.nationalgeographic.com/science/space/solar-system
http://solarsystem.nasa.gov/index.cfm

© **Educational Impressions, Inc.**

www.ingramcontent.com/pod-product-compliance
Lightning Source LLC
Chambersburg PA
CBHW051427200326

41520CB00023B/7382

* 9 7 8 1 5 6 6 4 4 0 5 0 9 *